Unstoppable Owl

Multiplication: **Pages:**

Multiplying Multiples of 10 1-4
Multiple Digits × 1 Digit 5-20
2 Digits × 2 Digits 21-27
Multiple Digits × Multiple Digits 28-33
Mixed Review 34-40

Division:

1-Digit Divisors 41-50
1-Digit Divisors with Remainders 51-60
1-Digit Divisors with Decimals 61-70
Multi-Digit Divisors 71-75
Multi-Digit Divisors with Remainders 76-80

Multiplication and Division:

Mixed Review 81-90
Word Problems 91-100

Answer Key 101-103

Name: Date: Score: Unstoppable Owl

1) Multiplying Multiples of 10 (with a Single-Digit Number)

When multiplying multiples of 10, focus on the **left-most** digit of each number, and for the time being, ignore the 0's. *If you multiplied the left-most digits, what would the answer be*? After that is determined, the 0's that were ignored can be written to the right.

Examples:	**Step 1**: Multiply (left-most)	**Step 2**: Write the 0's.	**Answer**:
70 × 5 =	7 × 5 = **35**	350	**350**
300 × 6 =	3 × 6 = **18**	1800	**1,800**
2 × 4,000 =	2 × 4 = **8**	8000	**8,000**
4 × 50 =	4 × 5 = **20**	200	**200**

Directions: *Solve each problem. Multiply the left-most digits. Then, write the zeros.*

1) **50 × 3 =**

2) **6 × 40 =**

3) **200 × 8 =**

4) **9 × 300 =**

5) **6,000 × 2 =**

6) **7 × 4,000 =**

7) **5 × 60 =**

8) **8 × 100 =**

9) **500 × 4 =**

10) **1,000 × 6 =**

11) **2,000 × 5 =**

12) **7 × 60 =**

Name: Date: Score: Unstoppable Owl

2) Multiplying Multiples of 10 (with a Single-Digit Number)

1) **8 × 600 =**

2) **30 × 7 =**

3) **9,000 × 2 =**

4) **400 × 4 =**

5) **30,000 × 4 =**

6) **6 × 100 =**

7) **800 × 5 =**

8) **6,000 × 6 =**

9) **900 × 8 =**

10) **2 × 70 =**

11) **2 × 50,000 =**

12) **700 × 9 =**

13) **50 × 8 =**

14) **4,000 × 2 =**

15) **7 × 700 =**

16) **200 × 8 =**

17) **9 × 10,000 =**

18) **300 × 6 =**

19) **9,000 × 9 =**

20) **4 × 80,000 =**

Name: Date: Score: Unstoppable Owl

3) Multiplying Multiples of 10 (with a Multiple of 10)

So far, you've practiced multiplying multiples of 10 by single-digit numbers. The same steps can be used when multiplying two multiples of 10 together. Focus on the **left-most** digit of each number. *If you multiplied the left-most digits, what would the answer be*? After that is determined, the 0's that were ignored can be written to the right.

When both numbers are a multiple of 10, both numbers will have 0's that are initially ignored. All of these zeros will need to be counted and written to the right.

Examples:	**Step 1**: Multiply (left-most)	**Step 2**: Write the 0's.	**Answer**:
70 × 50 =	7 × 5 = **35**	3500	**3,500**
600 × 400 =	6 × 4 = **24**	240000	**240,000**
30 × 3,000 =	3 × 3 = **9**	90000	**90,000**
5,000 × 800 =	5 × 8 = **40**	4000000	**4,000,000**

Directions: *Solve each problem. Multiply the left-most digits. Then, write the zeros.*

1) **60 × 70 =**

2) **200 × 30 =**

3) **800 × 400 =**

4) **20 × 7,000 =**

5) **600 × 500 =**

6) **50 × 4,000 =**

7) **100 × 100 =**

8) **3,000 × 400 =**

9) **40 × 400 =**

10) **10 × 90 =**

11) **7,000 × 30 =**

12) **1,000 × 60 =**

Name: Date: Score: Unstoppable Owl

4) Multiplying Multiples of 10

1) 300 × 500 =

2) 40 × 700 =

3) 5,000 × 200 =

4) 20 × 40 =

5) 9,000 × 60 =

6) 500 × 800 =

7) 600 × 60 =

8) 4,000 × 1,000 =

9) 70 × 80 =

10) 10 × 3,000 =

11) 20,000 × 60 =

12) 500 × 500 =

13) 100 × 100 =

14) 200 × 20 =

15) 40 × 6,000 =

16) 20 × 30 =

17) 60 × 800 =

18) 300 × 70 =

19) 50 × 60 =

20) 400 × 90,000 =

Name: Date: Score: Unstoppable Owl

5) Multiple Digits × 1 Digit

In this section, you will begin to learn how to multiply multi-digit numbers that are not multiples of 10 with a 1-digit number. One strategy to do this is by first writing the multi-digit number in expanded form. Then multiply each of the numbers in the expanded form with the single-digit number and add their products together.

Examples:	**Expanded Form:**	**Multiply:**	**Add to Find the Answer:**
72 × 5 =	70 + 2	70 × 5 = **350** 2 × 5 = **10**	**350** **+ 10** **360**
7 × 216 =	200 + 10 + 6 =	200 × 7 = **1,400** 10 × 7 = **70** 6 × 7 = **42**	**1,400** **70** **+ 42** **1,512**
3,908 × 2 =	3,000 + 900 + 0 + 8 =	3,000 × 2 = **6,000** 900 × 2 = **1,800** 0 × 2 = **0** 8 × 2 = **16**	**6,000** **1,800** **0** **+ 16** **7,816**

Directions: *Solve each problem.*

1) **45 × 8 =**

2) **3 × 91 =**

3) **271 × 4 =**

4) **5 × 632 =**

5) **83 × 9 =**

6) **6 × 227 =**

7) **1,048 × 7 =**

8) **2 × 9,253 =**

Name: Date: Score: Unstoppable Owl

6) Multiple Digits × 1 Digit

Writing the multi-digit number in expanded form is one strategy to multiply large numbers, but you don't have to write the expanded form. Many people multiply the 1-digit number and value of each of the digits in the multi-digit number. This accomplishes the same objective without writing the expanded form.

It does help to first write the problem **vertically**. Compare the examples below with the examples on page 5.

Examples:	**Rewritten Vertically**	**Multiply the Ones Column**	**Multiply the Tens Column**	**Multiply the Hundreds Column**	**Add**
72 × 5 =	72 × 5	72 × 5 **10** (2 × 5 = **10**)	72 × 5 10 **350** (70 × 5 = **350**)		72 × 5 10 +350 **360**
7 × 216 =	216 × 7	216 × 7 **42** (6 × 7 = **42**)	216 × 7 42 **70** (10 × 7 = **70**)	216 × 7 42 70 **1,400** (200 × 7 = **1,400**)	216 × 7 42 70 +1,400 **1,512**

Directions: *Solve each problem.*

1) **59 × 2**

2) **36 × 8**

3) **271 × 6**

4) **952 × 4**

5) **85 × 5**

6) **604 × 9**

7) **3,880 × 2**

8) **7,524 × 3**

Name: Date: Score: Unstoppable Owl

7) Multiple Digits × 1 Digit

1) 24 × 3

2) 326 × 8

3) 9,245 × 4

4) 581 × 7

5) 89 × 2

6) 6,622 × 6

7) 407 × 9

8) 18 × 5

9) 85 × 7

10) 792 × 4

11) 563 × 3

12) 28 × 6

13) 6,040 × 5

14) 16 × 8

15) 295 × 2

16) 38 × 9

17) 45 × 1

18) 5,495 × 7

19) 81 × 5

20) 673 × 3

Name: Date: Score: Unstopable Owl

8) Multiple Digits × 1 Digit

1) 117×3

2) 93×5

3) $4{,}872 \times 6$

4) 608×9

5) 429×7

6) $1{,}022 \times 4$

7) 739×2

8) 253×6

9) 939×8

10) 581×3

11) 632×8

12) $8{,}945 \times 4$

13) 72×5

14) 76×9

15) 94×2

16) 64×4

17) $21{,}573 \times 4$

18) $46{,}032 \times 3$

19) $94{,}225 \times 2$

20) $7{,}530 \times 5$

Name: Date: Score:

Unstoppable Owl

9) Multiple Digits × 1 Digit

1) 738×4

2) 525×2

3) $9{,}934 \times 6$

4) $5{,}087 \times 8$

5) 86×2

6) 79×7

7) 43×9

8) 61×3

9) 172×6

10) $3{,}900 \times 5$

11) 972×8

12) $7{,}847 \times 3$

13) 89×7

14) 314×4

15) $5{,}812 \times 9$

16) 76×5

17) $78{,}923 \times 6$

18) $10{,}572 \times 3$

19) $216{,}755 \times 5$

20) $8{,}199 \times 8$

Name: Date: Score: Unstoppable Owl

10) Multiple Digits × 1 Digit

1) 821 × 5

2) 156 × 3

3) 7,359 × 7

4) 6,408 × 2

5) 54 × 6

6) 99 × 9

7) 76 × 4

8) 58 × 8

9) 1,297 × 2

10) 573 × 9

11) 9,362 × 4

12) 623 × 7

13) 758 × 5

14) 65 × 3

15) 19 × 6

16) 2,973 × 9

17) 451,379 × 8

18) 84,223 × 7

19) 270,501 × 4

20) 15,600 × 3

Name: Date: Score: Unstoppable Owl

11) Multiple Digits × 1 Digit

1) 4,685 × 7

2) 527 × 5

3) 47 × 8

4) 1,214 × 6

5) 294 × 5

6) 5,897 × 2

7) 2,054 × 9

8) 386 × 3

9) 728 × 4

10) 84 × 7

11) 429 × 5

12) 9,339 × 8

13) 74 × 3

14) 28 × 9

15) 37 × 6

16) 165 × 2

17) 67,051 × 5

18) 752,934 × 8

19) 25,138 × 2

20) 7,832 × 9

Name: Date: Score: Unstopppable Owl

12) Multiple Digits × 1 Digit

So far, the problems in this section have been written vertically. The goal is for you to write the problems vertically yourself, and then solve them. Align the problems properly so that the digits in the ones column are positioned directly above one another. The same needs to be true for the digits in other columns.

Example 1: 23 × 5 =		**Example 2:** 8 × 417=	
Columns Aligned	**Columns Not Aligned**	**Columns Aligned**	**Columns Not Aligned**
$\begin{array}{r} 23 \\ \times\ \ 5 \\ \hline \end{array}$	$\begin{array}{r} 23 \\ \times 5\ \ \\ \hline \end{array}$	$\begin{array}{r} 417 \\ \times\ \ \ \ 8 \\ \hline \end{array}$	$\begin{array}{l} 417 \\ \times 8 \\ \hline \end{array}$

In the examples, the multi-digit number was written on the top. Many people prefer the vertical problems written this way, but they do not need to be. The 23 could have been written beneath the 5, and the 417 could have been written below the 8.

Regardless of which number you choose to write on the top, the steps for solving the problem are the same and the answer to the problem will be the same too.

Directions: *Rewrite each problem vertically, then solve.*

1) **45 × 8 =**

2) **3 × 91 =**

3) **271 × 4 =**

4) **5 × 632 =**

5) **83 × 9 =**

6) **6 × 227 =**

7) **1,048 × 7 =**

8) **2 × 9,253 =**

Name: Date: Score: Unstoppable Owl

13) Multiple Digits × 1 Digit

1) 526 × 7 =

2) 352 × 4 =

3) 8 × 79 =

4) 3 × 639 =

5) 6,275 × 5 =

6) 9 × 460 =

7) 817 × 6 =

8) 2 × 8,159 =

9) 53 × 7 =

10) 621 × 3 =

11) 8 × 95 =

12) 4 × 73 =

13) 14,528 × 2 =

14) 5 × 5,662 =

15) 23,084 × 7 =

16) 9 × 472 =

Name: Date: Score: *Unstoppable Owl*

14) Multiple Digits × 1 Digit

1) **4,762 × 5 =**

2) **638 × 8 =**

3) **3 × 214 =**

4) **6 × 1,992 =**

5) **7 × 68 =**

6) **471 × 9 =**

7) **2 × 985 =**

8) **555 × 4 =**

9) **2,472 × 6 =**

10) **794 × 3 =**

11) **86 × 8 =**

12) **5 × 349 =**

13) **2 × 526,042 =**

14) **41,819 × 4 =**

15) **7 × 12,986 =**

16) **9 × 3,824 =**

Name: Date: Score: Unstoppable Owl

15) Multiple Digits × 1 Digit

1) 6 × 450 =

2) 7,293 × 2 =

3) 8 × 917 =

4) 4 × 3,449 =

5) 3 × 105 =

6) 5 × 73 =

7) 416 × 7 =

8) 82 × 9 =

9) 2 × 759 =

10) 6,725 × 8 =

11) 572 × 5 =

12) 1,946 × 3 =

13) 5,931 × 6 =

14) 28,107 × 4 =

15) 9 × 94,522 =

16) 7 × 3,065 =

Name: Date: Score: Unstoppable Owl

16) Multiple Digits × 1 Digit

1) **67 × 4 =**

2) **7 × 58 =**

3) **73 × 9 =**

4) **5 × 84 =**

5) **3 × 5,821 =**

6) **472 × 8 =**

7) **894 × 2 =**

8) **9,253 × 6 =**

9) **618 × 3 =**

10) **9 × 405 =**

11) **6 × 2,594 =**

12) **8 × 57 =**

13) **46,827 × 5 =**

14) **3,704 × 7 =**

15) **4 × 13,028 =**

16) **2 × 7,992 =**

Name: Date: Score: Unstoppable Owl

17) Multiple Digits × 1 Digit

1) $8 \times 92 =$

2) $63 \times 5 =$

3) $7 \times 85 =$

4) $2 \times 345 =$

5) $580 \times 9 =$

6) $267 \times 3 =$

7) $6 \times 1{,}659 =$

8) $2{,}787 \times 4 =$

9) $5 \times 37 =$

10) $487 \times 8 =$

11) $6{,}853 \times 2 =$

12) $8{,}795 \times 6 =$

13) $7 \times 100{,}562 =$

14) $3 \times 57{,}389 =$

15) $9 \times 41{,}563 =$

16) $4 \times 9{,}522 =$

Name: Date: Score: Unstoppable Owl

18) Multiple Digits × 1 Digit

1) 2 × 5,517 =

2) 444 × 6 =

3) 7,035 × 9 =

4) 3 × 748 =

5) 48 × 5 =

6) 7 × 86 =

7) 92 × 4 =

8) 8 × 78 =

9) 8,116 × 3 =

10) 493 × 8 =

11) 754 × 2 =

12) 9 × 5,426 =

13) 30,065 × 7 =

14) 4 × 561,427 =

15) 5 × 61,200=

16) 6 × 4,736 =

Name: Date: Score: Unstoppable Owl

19) Multiple Digits × 1 Digit

1) $9 \times 99 =$

2) $28 \times 5 =$

3) $19 \times 7 =$

4) $3 \times 46 =$

5) $193 \times 8 =$

6) $4 \times 735 =$

7) $8,615 \times 2 =$

8) $6 \times 5,884 =$

9) $7 \times 9,500 =$

10) $624 \times 3 =$

11) $3,957 \times 9 =$

12) $5 \times 136 =$

13) $18,468 \times 4 =$

14) $2 \times 65,372 =$

15) $50,478 \times 6 =$

16) $8 \times 6,790 =$

Name: Date: Score:

Unstoppable Owl

20) Multiple Digits × 1 Digit

1) **83 × 6 =**

2) **2 × 77 =**

3) **9 × 32 =**

4) **45 × 3 =**

5) **7 × 485 =**

6) **4,936 × 4 =**

7) **257 × 8 =**

8) **5 × 681 =**

9) **684 × 2 =**

10) **4 × 8,492 =**

11) **119 × 7 =**

12) **6,315 × 9 =**

13) **8 × 52,748 =**

14) **11,539 × 5 =**

15) **3 × 218,634 =**

16) **6 × 5,714 =**

Name: Date: Score: Unstoppable Owl

21) 2 Digits × 2 Digits

Multiplying two 2-digit numbers is similar to the multiplication you've already done. Think of each number in expanded form, then multiply all the parts of the first number with all the parts of the second number.

Pay close attention to what each digit in the numbers represent. In the example below, the 6 in 64 does not represent 6. It represents 60 because it is in the tens place.

Example:	**Step 1: Multiply**	**Step 2: Multiply**	**Step 3: Multiply**	**Step 4: Multiply**	**Last Step: Add**
64 × 23 =	64 × 23 **12** (4 × 3 = **12**)	64 × 23 12 **180** (60 × 3 = **180**)	64 × 23 12 180 **80** (4 × 20 = **80**)	64 × 23 12 180 80 **1,200** (60 × 20 = **1,200**)	64 × 23 12 180 80 +1,200 1,472

Directions: *Solve each problem.*

1) **57 × 32**

2) **28 × 95**

3) **19 × 74**

4) **62 × 33**

5) **82 × 29**

6) **35 × 68**

7) **27 × 14**

8) **85 × 65**

Name: Date: Score: Unstoppable Owl

22) 2 Digits × 2 Digits

1) 62 × 78	2) 37 × 59	3) 55 × 81	4) 16 × 35
5) 93 × 46	6) 25 × 38	7) 67 × 82	8) 41 × 97
9) 54 × 57	10) 43 × 21	11) 13 × 15	12) 29 × 65
13) 50 × 64	14) 26 × 12	15) 38 × 34	16) 84 × 99

Name: Date: Score: Unstoppable Owl

23) 2 Digits × 2 Digits

1) 37×61

2) 28×45

3) 14×96

4) 88×44

5) 35×27

6) 19×24

7) 48×53

8) 90×75

9) 89×23

10) 77×18

11) 52×63

12) 26×17

13) 95×46

14) 67×39

15) 55×81

16) 60×92

Name: Date: Score: Unstoppable Owl

24) 2 Digits × 2 Digits

Directions: *Solve each problem.* ***It may help to write each problem vertically first.***

1) **43 × 27 =**

2) **86 × 91 =**

3) **13 × 17 =**

4) **22 × 54 =**

5) **94 × 38 =**

6) **65 × 78 =**

7) **32 × 26 =**

8) **70 × 45 =**

9) **56 × 42 =**

10) **39 × 74 =**

11) **15 × 40 =**

12) **28 × 21 =**

13) **97 × 89 =**

14) **49 × 27 =**

15) **46 × 18 =**

16) **78 × 24 =**

Name: Date: Score: Unstoppable Owl

25) 2 Digits × 2 Digits

1) **57 × 47 =**

2) **28 × 11 =**

3) **93 × 76 =**

4) **19 × 34 =**

5) **85 × 21 =**

6) **17 × 98 =**

7) **64 × 73 =**

8) **87 × 49 =**

9) **54 × 28 =**

10) **16 × 16 =**

11) **74 × 23 =**

12) **45 × 85 =**

13) **27 × 14 =**

14) **75 × 94 =**

15) **88 × 56 =**

16) **37 × 24 =**

Name: Date: Score: Unstoppable Owl

26) 2 Digits × 2 Digits

1) **81 × 35 =**

2) **27 × 98 =**

3) **14 × 62 =**

4) **36 × 24 =**

5) **95 × 61 =**

6) **80 × 33 =**

7) **63 × 48 =**

8) **17 × 19 =**

9) **89 × 25 =**

10) **37 × 42 =**

11) **55 × 22 =**

12) **74 × 98 =**

13) **16 × 79 =**

14) **54 × 86 =**

15) **27 × 51 =**

16) **48 × 59 =**

Name: Date: Score: Unstoppable Owl

27) 2 Digits × 2 Digits

1) 40 × 92 =

2) 31 × 56 =

3) 79 × 88 =

4) 23 × 73 =

5) 87 × 42 =

6) 30 × 37 =

7) 12 × 86 =

8) 34 × 45 =

9) 94 × 76 =

10) 18 × 59 =

11) 65 × 79 =

12) 90 × 83 =

13) 35 × 26 =

14) 68 × 21 =

15) 93 × 64 =

16) 72 × 45 =

Name: Date: Score: Unstoppable Owl

28) Multiple Digits × Multiple Digits

If you know how to multiply a 2-digit number by another 2-digit number, you have the skills to multiply larger numbers too. The same steps can be used. The main idea is to multiply the value of each digit from one number with the value of each digit in the other number.

Step 1: Multiply	Step 2: Multiply	Step 3: Multiply	Step 4: Multiply	Step 5: Multiply	Step 6: Multiply	Last Step: Add
319 × 52 **18**	319 × 52 18 **20**	319 × 52 18 20 **600**	319 × 52 18 20 600 **450**	319 × 52 18 20 600 450 **500**	319 × 52 18 20 600 450 500 **15,000**	319 × 52 18 20 600 450 500 + 15,000 16,588
(9 × 2)	(10 × 2)	(300 × 2)	(9 × 50)	(10 × 50)	(300 × 50)	

1) **627 × 83**

2) **259 × 41**

3) **368 × 95**

4) **174 × 24**

5) **556 × 27**

6) **703 × 49**

7) **216 × 32**

8) **944 × 87**

Name: Date: Score: Unstoppable Owl

29) Multiple Digits × Multiple Digits

Below are a couple more examples of multiplying multi-digit numbers. There is not enough room to show the individual steps, but the process of multiplying the value of each digit still applies. (Pay attention to the numbers in the parentheses to know which numbers were multiplied.)

Example 1: 235 × 461 =	**Example 2:** 3,568 × 24 =
235	3,568
× 461	× 24
5 ←(5 × 1)	32 ←(8 × 4)
30 ←(30 × 1)	240 ←(60 × 4)
200 ←(200 × 1)	2000 ←(500 × 4)
300 ←(5 × 60)	12,000 ←(3,000 × 4)
1,800 ←(30 × 60)	160 ←(8 × 20)
12,000 ←(200 × 60)	1,200 ←(60 × 20)
2,000 ←(5 × 400)	10,000 ←(500 × 20)
12,000 ←(30 × 400)	+ 60,000 ←(3,000 × 20)
+ 80,000 ←(200 × 400)	**85,632**
108,335	

Directions: *Solve each problem.*

1) **425 × 213**

2) **7,635 × 52**

3) **178 × 549**

4) **2,902 × 37**

Name: Date: Score: Unstoppable Owl

30) Multiple Digits × Multiple Digits

1) 2,739 × 61

2) 425 × 318

3) 957 × 42

4) 184 × 175

5) 950 × 436

6) 5,483 × 72

7) 4,700 × 35

8) 485 × 17

9) 694 × 27

10) 8,315 × 39

11) 14,642 × 23

12) 289 × 745

Name: Date: Score: Unstoppable Owl

31) Multiple Digits × Multiple Digits

1) 181×326

2) $4{,}837 \times 64$

3) 529×38

4) 630×512

5) $6{,}045 \times 97$

6) 984×13

7) 222×222

8) 681×375

9) 734×155

10) $3{,}271 \times 426$

11) 249×620

12) 384×90

Name: Date: Score: *Unstoppable Owl*

32) Multiple Digits × Multiple Digits

1) **276 × 415 =**

2) **683 × 29 =**

3) **17 × 1,846 =**

4) **91 × 3,358 =**

5) **204 × 376 =**

6) **75 × 2,653 =**

7) **29 × 857 =**

8) **243 × 51 =**

9) **483 × 726 =**

10) **4,294 × 17 =**

11) **85 × 7,223 =**

12) **471 × 295 =**

Name: Date: Score: Unstoppable Owl

33) Multiple Digits × Multiple Digits

1) **4,782 × 524 =**

2) **146 × 289 =**

3) **76 × 325 =**

4) **1,592 × 37 =**

5) **408 × 327 =**

6) **637 × 425 =**

7) **218 × 39 =**

8) **2,600 × 97 =**

9) **5,658 × 46 =**

10) **523 × 394 =**

11) **9,246 × 35 =**

12) **48 × 172 =**

Name: Date: Score: Unstoppable Owl

34) Mixed Review

1) 47 × 3,821 =

2) 294 × 216 =

3) 6,842 × 37 =

4) 5,763 × 2 =

5) 25 × 7,943 =

6) 462 × 935 =

7) 34,095 × 6 =

8) 14 × 2,865 =

9) 974 × 386 =

10) 24,361 × 15 =

11) 8 × 95,947 =

12) 34 × 875 =

Name: Date: Score: Unstoppable Owl

35) Mixed Review

1) **65 × 3,724 =**

2) **267 × 398 =**

3) **5 × 67,948 =**

4) **754 × 26 =**

5) **57 × 4,832 =**

6) **3 × 50,624 =**

7) **5,736 × 48 =**

8) **682 × 479 =**

9) **36,118 × 27 =**

10) **165,824 × 3 =**

11) **68 × 9,947 =**

12) **574 × 16 =**

Name: Date: Score: Unstoppable Owl

36) Mixed Review

1) 3,781 × 4 =

2) 72 × 45 =

3) 293 × 516 =

4) 8,290 × 32 =

5) 57 × 653 =

6) 3 × 19,625 =

7) 5,400 × 24 =

8) 9 × 2,078 =

9) 6,272 × 15 =

10) 328 × 700 =

11) 461,593 × 2 =

12) 95 × 64 =

Name: Date: Score:

37) Mixed Review

1) **472 × 162 =**

2) **79 × 463 =**

3) **3,882 × 29 =**

4) **8 × 7,214 =**

5) **50,473 × 4 =**

6) **36 × 92 =**

7) **289 × 45 =**

8) **1,348 × 55 =**

9) **729 × 615 =**

10) **42 × 6,551 =**

11) **3 × 65,834 =**

12) **14 × 958 =**

Name: Date: Score: Unstoppable Owl

38) Mixed Review

1) $75 \times 3,872 =$

2) $539 \times 40 =$

3) $17 \times 9,846 =$

4) $73 \times 6,808 =$

5) $5 \times 89,421 =$

6) $26 \times 99 =$

7) $4,968 \times 37 =$

8) $22 \times 7,481 =$

9) $6 \times 71,928 =$

10) $200,560 \times 4 =$

11) $947 \times 365 =$

12) $128 \times 84 =$

Name: Date: Score: Unstoppable Owl

39) Mixed Review

1) **15,742 × 8 =**

2) **45 × 2,860 =**

3) **483 × 297 =**

4) **176 × 38 =**

5) **14 × 9,803 =**

6) **2 × 478,194 =**

7) **63 × 324 =**

8) **16 × 18 =**

9) **507 × 394 =**

10) **29 × 4,392 =**

11) **312,954 × 3 =**

12) **81 × 77 =**

Name: Date: Score: Unstoppable Owl

40) Mixed Review

1) 59 × 38 =

2) 360 × 283 =

3) 4,178 × 42 =

4) 7 × 6,154 =

5) 3,724 × 53 =

6) 9 × 967 =

7) 628 × 55 =

8) 6 × 1,893 =

9) 24,700 × 23 =

10) 739 × 446 =

11) 5 × 34,835 =

12) 412 × 89 =

Name: Date: Score: Unstoppable Owl

41) 1-Digit Divisors

There are several ways to divide multi-digit numbers. Two common methods are the **partial products method** and the **long division method**. Both will be introduced in this book.

The partial products method usually has **many ways to solve a division problem.** The examples below show the same problem being solved in two different ways.

To use the partial products method, think of a number you can multiply by the **divisor** that will result in a product that is less than the number being divided. In the example, *what could you multiply by 2 that would be less than 174*? Example 1 multiplied by 50. Example 2 multiplied by 80. Either one of these is a good number to use. When multiplied by 2, they create a product that is less than 174. After subtracting, another number needs to be multiplied by the divisor to find a product that is less than the remaining number. In example 1, the remaining number is 74. In example 2, the remaining number is 14.

Keep finding partial products until the number remaining is 0. At that point, add the multiples that were multiplied by 2. The result will be the answer to the original problem.

Example 1: 174 ÷ 2 =	**Example 2:** 174 ÷ 2 =
2)174 − 100 ← (2 × **50** = 100) **Step 1:** 74 **Step 2:** − 60 ← (2 × **30** = 60) **Step 3:** 14 **Step 4:** − 14 ← (2 × **7** = 14) **Step 5:** 0 + **Step 6:** **87** **Answer:** 174 ÷ 2 = **87**	2)174 − 160 ← (2 × **80** = 160) **Step 1:** 14 **Step 2:** − 14 ← (2 × **7** = 14) **Step 3:** 0 + **Step 4:** **87** **Answer:** 174 ÷ 2 = **87**
Step 1: I know 2 × 50 is 100. **Step 2:** 174 − 100 = 74 **Step 3:** I know 2 × 30 is 60. **Step 4:** 74 − 60 = 14 **Step 5:** I know 2 × 7 is 14. **Step 6:** 14 − 14 = 0. Add the multiples. 50 + 30 + 7 = **87**. **Answer:** 174 ÷ 2 = **87**	**Step 1:** I know 2 × 80 is 160. **Step 2:** 174 − 160 = 14 **Step 3:** I know 2 × 7 is 14. **Step 4:** 14 − 14 = 0. Add the multiples. 50 + 30 + 7 = **87**. **Answer:** 174 ÷ 2 = **87**

1) **2)58** 2) **3)126** 3) **5)85** 4) **4)72**

Name: Date: Score: Unstoppable Owl

42) 1-Digit Divisors

Long division is more systematic and only has one way to solve the problem.

First, look at the left-most digit of the **dividend**. *How many times can the divisor fit into the left-most digit of the dividend*? In the example, the dividend is 174 and the left-most digit is 1. *How many times does 2 go into 1*? 2 is bigger so it cannot fit into 1.

Now look at the number beside the 1. Together, the 1 and 7 form a 17. *How many times can 2 go into 17*? 2 can go into 17 a total of 8 times, so an 8 can be written above the 17. 8 × 2 = 16, and 17 – 16 = 1.

Now, the next digit of the dividend needs to be moved down next to this 1. Together the 1 and the 4 form 14. How many times can 2 go into 14? 2 can go into 14 a total of 7 times, so write 7 above the 4. 2 × 7 = 14, and 14 – 14 = 0. There are no more numbers to divide.

The numbers above form the answer. 174 ÷ 2 = **87**.

Example: 174 ÷ 2 =					
Step 1:	**Step 2:**	**Step 3:**	**Step 4:**	**Step 5:**	**Step 6:**
2) **174**	**8** 2) **174**	8 2) 174 – 16 **1**	8 2) 174 – 16 **14**	**87** 2) 174 – 16 14	(87) 2) 174 – 16 14 – 14 **0**

Step 1: *How many times does 2 go into 1*? It can't, move to step 2.

Step 2: *How many times does 2 go into 17*? 8 times. So, write 8 above the 17.

Step 3: 2 × 8 is 16. Subtract 16 from the 17. 17 – 16 = 1.

Step 4: Bring down then next number (which is 4) and write it next to the 1.

Step 5: How many times does 2 go into 14? 7 times. So, write 7 above the 4.

Step 6: 2 × 7 is 14. Subtract 14 from the 14. 14 – 14 = 0. The division is complete.

Answer: 174 ÷ 2 = **87**

1) **2) 86**

2) **5) 625**

3) **4) 112**

4) **3) 168**

Name: Date: Score: Unstoppable Owl

43) 1-Digit Divisors

Partial products or long division? *Which method do you prefer*? Practice them both.
Decide which method you like best and use it to complete the problems in this section.

1) $5\overline{)170}$

2) $3\overline{)771}$

3) $4\overline{)324}$

4) $2\overline{)366}$

5) $7\overline{)161}$

6) $6\overline{)858}$

7) $3\overline{)177}$

8) $5\overline{)955}$

9) $2\overline{)3,516}$

10) $3\overline{)3,216}$

11) $4\overline{)9,128}$

12) $4\overline{)860}$

Unstoppable Owl

Name: Date: Score:

44) 1-Digit Divisors

1) $2\overline{)94}$

2) $5\overline{)3{,}185}$

3) $6\overline{)1{,}284}$

4) $3\overline{)873}$

5) $7\overline{)252}$

6) $2\overline{)922}$

7) $4\overline{)372}$

8) $8\overline{)936}$

9) $2\overline{)45{,}874}$

10) $3\overline{)10{,}524}$

11) $6\overline{)9{,}522}$

12) $5\overline{)265}$

Name: Date: Score: Unstoppable Owl

45) 1-Digit Divisors

1) $6\overline{)2{,}454}$

2) $4\overline{)208}$

3) $2\overline{)1{,}470}$

4) $5\overline{)445}$

5) $3\overline{)87}$

6) $7\overline{)4{,}690}$

7) $5\overline{)1{,}810}$

8) $9\overline{)2{,}187}$

9) $4\overline{)1{,}592}$

10) $3\overline{)37{,}806}$

11) $2\overline{)7{,}944}$

12) $8\overline{)6{,}000}$

Name: Date: Score: Unstoppable Owl

46) 1-Digit Divisors

1) $4\overline{)2,292}$

2) $8\overline{)792}$

3) $6\overline{)1,872}$

4) $3\overline{)9,144}$

5) $5\overline{)1,465}$

6) $5\overline{)3,595}$

7) $2\overline{)7,508}$

8) $4\overline{)304}$

9) $4\overline{)20,672}$

10) $5\overline{)11,850}$

11) $2\overline{)67,498}$

12) $7\overline{)581}$

Name: Date: Score: Unstoppable Owl

47) 1-Digit Divisors

1) $2\overline{)680}$

2) $8\overline{)6,328}$

3) $3\overline{)1,626}$

4) $4\overline{)3,824}$

5) $9\overline{)513}$

6) $5\overline{)3,580}$

7) $9\overline{)1,557}$

8) $7\overline{)3,108}$

9) $5\overline{)19,755}$

10) $4\overline{)34,120}$

11) $3\overline{)8,247}$

12) $4\overline{)712}$

Name: Date: Score: Unstoppable Owl

48) 1-Digit Divisors

1) $9\overline{)4{,}212}$

2) $7\overline{)644}$

3) $4\overline{)72}$

4) $3\overline{)7{,}812}$

5) $5\overline{)1{,}130}$

6) $2\overline{)1{,}936}$

7) $3\overline{)1{,}905}$

8) $3\overline{)8{,}526}$

9) $6\overline{)27{,}000}$

10) $4\overline{)5{,}000}$

11) $3\overline{)75{,}315}$

12) $5\overline{)3{,}220}$

Name: Date: Score: Unstoppable Owl

49) 1-Digit Divisors

1) **879 ÷ 3 =**

2) **1,328 ÷ 4 =**

3) **910 ÷ 5 =**

4) **4,762 ÷ 2 =**

5) **846 ÷ 6 =**

6) **8,700 ÷ 4 =**

7) **6,170 ÷ 2 =**

8) **5,776 ÷ 8 =**

9) **9,651 ÷ 3 =**

10) **36,492 ÷ 6 =**

11) **18,630 ÷ 5 =**

12) **2,652 ÷ 4 =**

Name: Date: Score: Unstoppable Owl

50) 1-Digit Divisors

1) 756 ÷ 3 =

2) 74 ÷ 2 =

3) 3,506 ÷ 2 =

4) 4,700 ÷ 5 =

5) 1,001 ÷ 7 =

6) 894 ÷ 3 =

7) 6,684 ÷ 4 =

8) 5,000 ÷ 8 =

9) 14,338 ÷ 2 =

10) 10,500 ÷ 6 =

11) 41,975 ÷ 5 =

12) 6,112 ÷ 8 =

Name: Date: Score: *Unstoppable Owl*

51) 1-Digit Divisors with Remainders

Pretend a group of 4 people ordered a pizza. If the pizza has 8 slices, how many slices will each person get if they divide the slices evenly? 8 slices ÷ 4 people = **2 slices per person**.

Now pretend there were only 3 people, and they decide to order a pizza. If the pizza has 8 slices, how many slices will each person get if they divide the slices evenly? What is 8 slices ÷ 3 people? Since 8 can't be divided by 3 evenly, each person can have 2 slices, but there will be 2 slices left over.

When dividing, these *leftovers* are called **remainders**. You can say that the answer to 8 ÷ 3 is **2 remainder 2**. This can also be abbreviated by writing **2 R2** (the R stands for remainder).

Example 1: 8 slices ÷ 4 people = **2 slices per person.**

Person 1 Person 2 Person 3 Person 4

Example 2: 8 slices ÷ 3 people = **2 slices per person with 2 slices left over** (2 R2)

Person 1 Person 2 Person 3

Leftovers

Directions: *Solve each problem and state if there are any remainders. Draw a picture if it helps.*

1) **There are 15 apples and 2 buckets.**
If divided evenly between the buckets,
how many apples will be in each bucket?
Are there any remainders?

2) **There are 19 coins and 4 kids.**
If divided evenly among themselves,
how many coins will each kid get?
Are there any remainders?

3) **There are 15 cookies and 5 people.**
If divided evenly between the people,
how many cookies will each person get?
Are there any remainders?

4) **There are 20 carrots and 3 rabbits.**
If divided evenly between the rabbits,
how many carrots will each rabbit get?
Are there any remainders?

Name: Date: Score:

52) 1-Digit Divisors with Remainders

Drawing pictures may help solve simple division problems with remainders, but the partial products method and long division method are better suited for solving more difficult division problems. These methods have already been introduced. The only difference is that now there may be remainders after all the division is complete.

The examples below show the same problem being solved. One example shows the problem being solved using the partial products method. The other example shows the problem being solved using long division.

By now, you may prefer one method over the other. That's fine. Use whichever method you like to solve the problems in this section.

Partial Products Example: 573 ÷ 2 =	**Long Division Example:** 573 ÷ 2 =
2)573 − 500 ← (2 × **250** = 500) **Step 1:** 73 **Step 2:** − 60 ← (2 × **30** = 60) **Step 3:** 13 **Step 4:** − 12 ← (2 × **6** = 12) **Step 5:** 1 + **Step 6:** 286 **Answer:** 573 ÷ 2 = **286 R1**	286 2)573 − 4 17 − 16 13 − 12 1 573 ÷ 2 = **286 R1**

Directions: *Solve each problem and identify any remainders.*

1) **2)75**

2) **3)176**

3) **5)684**

4) **9)1,985**

Name: Date: Score: Unstoppable Owl

53) 1-Digit Divisors with Remainders

Directions: *Solve each problem and identify any remainders.*

1) $5\overline{)84}$

2) $7\overline{)408}$

3) $6\overline{)7,538}$

4) $9\overline{)143}$

5) $6\overline{)791}$

6) $2\overline{)3,219}$

7) $4\overline{)5,542}$

8) $3\overline{)1,384}$

9) $7\overline{)25,000}$

10) $8\overline{)12,548}$

11) $9\overline{)7,046}$

12) $6\overline{)460}$

Name: Date: Score: Unstoppable Owl

54) 1-Digit Divisors with Remainders

Directions: *Solve each problem and identify any remainders.*

1) $5\overline{)623}$

2) $3\overline{)1,040}$

3) $2\overline{)91}$

4) $6\overline{)1,937}$

5) $4\overline{)756}$

6) $5\overline{)3,179}$

7) $7\overline{)6,815}$

8) $9\overline{)2,000}$

9) $3\overline{)14,827}$

10) $5\overline{)9,548}$

11) $4\overline{)7,832}$

12) $6\overline{)3,800}$

Name: Date: Score: Unstoppable Owl

55) 1-Digit Divisors with Remainders

Directions: *Solve each problem and identify any remainders.*

1) $8\overline{)7,558}$

2) $7\overline{)642}$

3) $6\overline{)8,000}$

4) $5\overline{)629}$

5) $4\overline{)5,421}$

6) $9\overline{)4,250}$

7) $8\overline{)999}$

8) $4\overline{)137}$

9) $2\overline{)6,645}$

10) $6\overline{)50,837}$

11) $9\overline{)16,740}$

12) $3\overline{)2,834}$

Name: Date: Score: Unstoppable Owl

56) 1-Digit Divisors with Remainders

1) 5,730 ÷ 4 =

2) 677 ÷ 2 =

3) 3,000 ÷ 7 =

4) 1,285 ÷ 6 =

5) 756 ÷ 3 =

6) 2,815 ÷ 8 =

7) 271 ÷ 4 =

8) 4,093 ÷ 5 =

9) 13,579 ÷ 5 =

10) 11,934 ÷ 4 =

11) 6,722 ÷ 3 =

12) 26,750 ÷ 9 =

Name: Date: Score:

Unstoppable Owl

57) 1-Digit Divisors with Remainders

1) 3,488 ÷ 6 =

2) 400 ÷ 9 =

3) 2,197 ÷ 5 =

4) 759 ÷ 8 =

5) 8,854 ÷ 7 =

6) 625 ÷ 4 =

7) 1,500 ÷ 6 =

8) 4,879 ÷ 5 =

9) 61,842 ÷ 5 =

10) 15,649 ÷ 8 =

11) 2,300 ÷ 3 =

12) 4,073 ÷ 6 =

Name: Date: Score: Unstoppable Owl

58) 1-Digit Divisors with Remainders

1) **7,275 ÷ 9 =**

2) **5,611 ÷ 6 =**

3) **882 ÷ 7 =**

4) **1,010 ÷ 3 =**

5) **3,904 ÷ 5 =**

6) **1,535 ÷ 4 =**

7) **952 ÷ 6 =**

8) **5,299 ÷ 8 =**

9) **95,102 ÷ 6 =**

10) **41,772 ÷ 5 =**

11) **794 ÷ 9 =**

12) **2,311 ÷ 3 =**

Name: Date: Score: Unstoppable Owl

59) 1-Digit Divisors with Remainders

1) 2,784 ÷ 5 =

2) 548 ÷ 3 =

3) 1,122 ÷ 8 =

4) 9,310 ÷ 4 =

5) 6,422 ÷ 7 =

6) 2,570 ÷ 6 =

7) 639 ÷ 2 =

8) 1,950 ÷ 3 =

9) 12,847 ÷ 4 =

10) 8,521 ÷ 9 =

11) 27,385 ÷ 3 =

12) 7,592 ÷ 5 =

Name: Date: Score: Unstoppable Owl

60) 1-Digit Divisors with Remainders

1) **479 ÷ 3 =**

2) **2,680 ÷ 6 =**

3) **1,195 ÷ 7 =**

4) **5,072 ÷ 6 =**

5) **853 ÷ 4 =**

6) **7,561 ÷ 3 =**

7) **2,994 ÷ 8 =**

8) **680 ÷ 9 =**

9) **7,155 ÷ 8 =**

10) **36,317 ÷ 5 =**

11) **15,842 ÷ 7 =**

12) **5,924 ÷ 6 =**

Name: Date: Score: *Unstoppable Owl*

61) 1-Digit Divisors with Decimals

Amounts that are left over can be written as remainders, but they can also be calculated as a decimal instead. The long division method is often easier for calculating decimals.

The steps for dividing are the same until the remainder is found. Instead of stopping the division problem, a decimal point and a zero can be written behind the number being divided. This zero can be moved down next to the remainder. Now, an additional step of division can take place.

Since a decimal point was added behind the dividend, a decimal point directly above it needs to be added in the answer.

Example: 86 ÷ 4 =

Step 1:	**Step 2:**	**Step 3:**	**Step 4:**	**Step 5:**	**Step 6:**
21 4)86 − 8 06 − 4 2	21 4)86.0 − 8 06 − 4 2	21. 4)86.0 − 8 06 − 4 2	21. 4)86.0 − 8 06 − 4 20	21.5 4)86.0 − 8 06 − 4 20	(21.5) 4)86.0 − 8 06 − 4 20 − 20 0

Step 1: Divide normally until you find the remainder.
Step 2: Write a decimal point and a zero behind the dividend.
Step 3: Write a decimal point behind the answer that has been calculated so far.
Step 4: Bring down the zero next to the remainder.
Step 5: *How many times does 4 go into 20*? 5 times. So, write 5 above the 0.
Step 6: 4 × 5 is 20. Subtract 20 from the 20. 20 – 20 = 0. The division is complete.
Answer: 86 ÷ 4 = **21.5**

Directions: *Solve each division problem. When not a whole number, state the answer as a decimal.*

1) **5)73** 2) **2)235** 3) **4)638** 4) **5)374**

Name: Date: Score: Unstoppable Owl

62) 1-Digit Divisors with Decimals

Directions: *Solve each division problem. When not a whole number, state the answer as a decimal.*

1) $2\overline{)57}$

2) $4\overline{)90}$

3) $5\overline{)3,196}$

4) $4\overline{)7,930}$

5) $5\overline{)839}$

6) $6\overline{)81}$

7) $2\overline{)1,725}$

8) $5\overline{)3,447}$

9) $5\overline{)10,058}$

10) $4\overline{)7,198}$

11) $5\overline{)76,271}$

12) $2\overline{)6,839}$

Name: Date: Score: *Unstoppable Owl*

63) 1-Digit Divisors with Decimals

So far, the problems with decimals have had answers with one digit after the decimal. This digit is in the **tenths** place. Today, the answers may have more digits after the decimal. Each time a remainder is found, write a decimal then continue writing zeros and dividing until there is no longer a remainder.

Example 1: $29 \div 8 =$	**Example 2:** $75 \div 4 =$	**Example 3:** $123 \div 8 =$	**Example 4:** $21 \div 4 =$
3.625 8) 29.000 − 24 50 − 48 20 − 16 40 − 40 0	**18.75** 4) 75.00 − 4 35 − 32 30 − 28 20 − 20 0	**15.375** 8) 123.000 − 8 43 − 40 30 − 24 60 − 56 40 − 40 0	**5.25** 4) 21.00 − 20 10 − 8 20 − 20 0
Answer 1: $29 \div 8 =$ **3.625**	**Answer 2:** $75 \div 4 =$ **18.75**	**Answer 3:** $123 \div 8 =$ **15.375**	**Answer 4:** $21 \div 4 =$ **5.25**

Directions: *Solve each division problem. When not a whole number, state the answer as a decimal.*

1) **4) 81**

2) **4) 11**

3) **8) 317**

4) **8) 254**

Name: Date: Score:

64) 1-Digit Divisors with Decimals

Directions: *Solve each division problem. When not a whole number, state the answer as a decimal.*

1) $8\overline{)165}$

2) $4\overline{)291}$

3) $6\overline{)345}$

4) $5\overline{)843}$

5) $4\overline{)137}$

6) $8\overline{)263}$

7) $4\overline{)350}$

8) $8\overline{)550}$

9) $2\overline{)25}$

10) $4\overline{)3}$

11) $8\overline{)625}$

12) $4\overline{)111}$

Name: Date: Score: Unstoppable Owl

65) 1-Digit Divisors with Decimals

You can continue dividing problems and calculate long decimals. It just takes a while. But sometimes, when dividing and calculating a decimal, the division never ends. No matter how many decimal places are calculated, there will always be a remainder left to divide.

Most of the time, you may not need a completely accurate answer. If you calculate the division problem a few decimal places out, then round, the **rounded answer** is accurate enough. The rest of this section will focus on rounding answers to the nearest **hundredths** place (two place values behind the decimal). To do this, calculate an answer to the **thousandths** place. Then, stop and use that digit to round the hundredths place up or down.

Examples:	**Answers:**	**Rounded Answers:** (to the Nearest Hundredth)
$67 \div 9 =$	7.4444444...	7.44
$118 \div 7 =$	16.857142...	16.86
$51 \div 8 =$	6.375	6.38
$547 \div 3 =$	182.3333333...	182.33
$16 \div 6 =$	2.6666666...	2.67

Directions: *Round each number* ***to the nearest hundredth****.*

1) **7.3817**
2) **25.06724**
3) **1.9549**
4) **214.3333**
5) **5.666666**
6) **8.71593**
7) **40.7846**
8) **19.997**
9) **143.2833**
10) **10.252525**
11) **0.3984**
12) **637.805**

Directions: *Solve each division problem. If the answer has at least three decimal places, stop dividing and* ***round the answer to the nearest hundredth****.*

13) $9\overline{)74}$
14) $6\overline{)145}$
15) $3\overline{)88}$
16) $8\overline{)17}$

Name: Date: Score: Unstoppable Owl

66) 1-Digit Divisors with Decimals

Directions: *Solve each division problem. If the answer has at least three decimal places, stop dividing and* ***round the answer to the nearest <u>hundredth</u>***.

1) $8\overline{)355}$ 2) $9\overline{)584}$ 3) $7\overline{)492}$ 4) $3\overline{)278}$

5) $9\overline{)52}$ 6) $3\overline{)43}$ 7) $6\overline{)265}$ 8) $9\overline{)1,463}$

9) $8\overline{)121}$ 10) $6\overline{)470}$ 11) $9\overline{)250}$ 12) $6\overline{)299}$

Name: Date: Score: Unstoppable Owl

67) 1-Digit Divisors with Decimals

Directions: *Solve each division problem. If the answer has at least three decimal places, stop dividing and* ***round the answer to the nearest <u>hundredth</u>****.*

1) $9\overline{)736}$

2) $7\overline{)549}$

3) $6\overline{)95}$

4) $3\overline{)164}$

5) $7\overline{)247}$

6) $6\overline{)511}$

7) $8\overline{)13}$

8) $7\overline{)27}$

9) $4\overline{)50}$

10) $3\overline{)758}$

11) $8\overline{)327}$

12) $9\overline{)5}$

Name: Date: Score: Unstoppable Owl

68) 1-Digit Divisors with Decimals

Directions: *Solve each division problem. If the answer has at least three decimal places, stop dividing and* ***round the answer to the nearest <u>hundredth</u>***.

1) $3\overline{)284}$

2) $8\overline{)571}$

3) $9\overline{)37}$

4) $7\overline{)688}$

5) $6\overline{)59}$

6) $7\overline{)153}$

7) $9\overline{)386}$

8) $8\overline{)94}$

9) $3\overline{)859}$

10) $9\overline{)214}$

11) $8\overline{)589}$

12) $3\overline{)71}$

Name: Date: Score: *Unstoppable Owl*

69) 1-Digit Divisors with Decimals

Directions: *Solve each division problem. If the answer has at least three decimal places, stop dividing and* ***round the answer to the nearest <u>hundredth</u>***.

1) $7\overline{)445}$ 2) $7\overline{)57}$ 3) $9\overline{)26}$ 4) $9\overline{)293}$

5) $6\overline{)82}$ 6) $6\overline{)952}$ 7) $8\overline{)77}$ 8) $7\overline{)25}$

9) $8\overline{)399}$ 10) $6\overline{)175}$ 11) $4\overline{)259}$ 12) $6\overline{)413}$

Name: Date: Score: Unstoppable Owl

70) 1-Digit Divisors with Decimals

Directions: *Solve each division problem. If the answer has at least three decimal places, stop dividing and* ***round the answer to the nearest <u>hundredth</u>****.*

1) $6\overline{)329}$

2) $8\overline{)3}$

3) $9\overline{)177}$

4) $8\overline{)63}$

5) $7\overline{)680}$

6) $3\overline{)251}$

7) $4\overline{)139}$

8) $8\overline{)43}$

9) $7\overline{)3,482}$

10) $5\overline{)473}$

11) $9\overline{)488}$

12) $8\overline{)165}$

Name: Date: Score: Unstoppable Owl

71) Multi-Digit Divisors

The same steps used for solving division problems with single-digit divisors can be used to solve problems with multi-digit divisors. You may use the partial products method or long division method to solve these problems too.

Partial Products Example: 645 ÷ 15 =	**Long Division Example:** 645 ÷ 15 =
15) 645 − 300 ← (15 × **20** = 300) **Step 1:** 345 **Step 2:** − 300 ← (15 × **20** = 300) **Step 3:** 45 **Step 4:** − 45 ← (15 × **3** = 45) **Step 5:** 0 **+** **Step 6:** (**43**) **Answer:** 645 ÷ 15 = **43**	(**43**) 15) 645 − 60 45 − 45 **0** 645 ÷ 15 = **43**

1) **15) 540**

2) **12) 1,776**

3) **25) 875**

4) **11) 946**

5) **30) 1,950**

6) **13) 377**

7) **21) 3,297**

8) **12) 2,772**

Name: Date: Score: Unstoppable Owl

72) Multi-Digit Divisors

1) $11\overline{)5,082}$

2) $22\overline{)5,346}$

3) $25\overline{)10,275}$

4) $15\overline{)495}$

5) $21\overline{)11,046}$

6) $12\overline{)6,240}$

7) $13\overline{)312}$

8) $20\overline{)760}$

9) $15\overline{)2,145}$

10) $11\overline{)4,202}$

11) $20\overline{)16,420}$

12) $12\overline{)5,076}$

Name: Date: Score: *Unstoppable Owl*

73) Multi-Digit Divisors

1) $12\overline{)3{,}648}$

2) $31\overline{)651}$

3) $40\overline{)8{,}600}$

4) $22\overline{)7{,}524}$

5) $20\overline{)12{,}860}$

6) $13\overline{)2{,}639}$

7) $21\overline{)8{,}652}$

8) $15\overline{)9{,}780}$

9) $11\overline{)8{,}019}$

10) $25\overline{)35{,}775}$

11) $12\overline{)29{,}208}$

12) $30\overline{)82{,}620}$

Name: Date: Score: Unstopable Owl

74) Multi-Digit Divisors

1) $32\overline{)1{,}344}$

2) $11\overline{)7{,}469}$

3) $13\overline{)6{,}812}$

4) $25\overline{)18{,}700}$

5) $200\overline{)6{,}800}$

6) $15\overline{)3{,}600}$

7) $12\overline{)49{,}620}$

8) $30\overline{)38{,}250}$

9) $50\overline{)43{,}350}$

10) $12\overline{)28{,}836}$

11) $150\overline{)64{,}650}$

12) $120\overline{)64{,}080}$

Name: Date: Score:

Unstoppable Owl

75) Multi-Digit Divisors

1) $300\overline{)162{,}600}$

2) $25\overline{)16{,}200}$

3) $12\overline{)40{,}944}$

4) $110\overline{)28{,}270}$

5) $250\overline{)81{,}000}$

6) $20\overline{)18{,}760}$

7) $200\overline{)150{,}600}$

8) $50\overline{)48{,}200}$

9) $11\overline{)8{,}019}$

10) $21\overline{)86{,}373}$

11) $120\overline{)5{,}400}$

12) $102\overline{)43{,}248}$

Name: Date: Score: Unstoppable Owl

76) Multi-Digit Divisors with Remainders

Once you know how to solve problems with multi-digit divisors without remainders, you can solve problems that have remainders too. Follow the same steps as before. Divide the problem as much as you can. If some of the amount is left over, it's the remainder.

Partial Products Example: 781 ÷ 25 =	**Long Division Example:** 781 ÷ 25 =
25)781 − 500 ← (25 × **20** = 500) **Step 1:** 281 **Step 2:** − 250 ← (25 × **10** = 250) **Step 3:** 31 **Step 4:** − 25 ← (25 × **1** = 25) **Step 5:** (6) + **Step 6:** (31) **Answer:** 781 ÷ 25 = **31 R6**	(31) 25)781 − 75 31 − 25 (6) 781 ÷ 25 = **31 R6**

1) **15)627**

2) **20)4,165**

3) **11)894**

4) **12)2,314**

5) **30)826**

6) **21)1,200**

7) **13)425**

8) **25)3,795**

Name: Date: Score: Unstoppable Owl

77) Multi-Digit Divisors with Remainders

Directions: *Solve each problem and identify any remainders.*

1) $12\overline{)5{,}139}$

2) $25\overline{)13{,}872}$

3) $40\overline{)94{,}732}$

4) $15\overline{)593}$

5) $21\overline{)6{,}494}$

6) $13\overline{)4{,}675}$

7) $22\overline{)89{,}653}$

8) $50\overline{)42{,}839}$

9) $20\overline{)17{,}245}$

10) $15\overline{)7{,}986}$

11) $11\overline{)10{,}360}$

12) $30\overline{)2{,}172}$

Name: Date: Score: Unstoppable Owl

78) Multi-Digit Divisors with Remainders

Directions: *Solve each problem and identify any remainders.*

1) $50\overline{)35{,}116}$

2) $11\overline{)3{,}720}$

3) $32\overline{)6{,}850}$

4) $25\overline{)13{,}618}$

5) $12\overline{)4{,}915}$

6) $20\overline{)7{,}053}$

7) $15\overline{)8{,}119}$

8) $30\overline{)852}$

9) $25\overline{)1{,}784}$

10) $40\overline{)5{,}526}$

11) $21\overline{)15{,}187}$

12) $13\overline{)4{,}470}$

Name: Date: Score: Unstoppable Owl

79) Multi-Digit Divisors with Remainders

Directions: *Solve each problem and identify any remainders.*

1) $30\overline{)9{,}487}$

2) $80\overline{)2{,}539}$

3) $12\overline{)5{,}000}$

4) $25\overline{)8{,}195}$

5) $11\overline{)684}$

6) $200\overline{)48{,}355}$

7) $15\overline{)1{,}972}$

8) $50\overline{)12{,}984}$

9) $150\overline{)8{,}437}$

10) $40\overline{)5{,}110}$

11) $20\overline{)1{,}836}$

12) $23\overline{)740}$

Name: Date: Score: Unstoppable Owl

80) Multi-Digit Divisors with Remainders

Directions: *Solve each problem and identify any remainders.*

1) $21\overline{)3{,}000}$

2) $250\overline{)7{,}847}$

3) $15\overline{)2{,}222}$

4) $40\overline{)25{,}704}$

5) $25\overline{)11{,}648}$

6) $12\overline{)500}$

7) $30\overline{)8{,}628}$

8) $200\overline{)5{,}783}$

9) $15\overline{)26{,}736}$

10) $300\overline{)2{,}756}$

11) $20\overline{)1{,}460}$

12) $50\overline{)9{,}468}$

Name: Date: Score: Unstoppable Owl

81) Mixed Review

Directions: *Solve each problem and identify any remainders.*

1) **527 × 8 =**

2) **45 × 62 =**

3) **952 × 35 =**

4) **6 × 8,732 =**

5) **155 × 29 =**

6) **67 × 40 =**

7) **754 ÷ 4 =**

8) **2,675 ÷ 5 =**

9) **3,582 ÷ 20 =**

10) **973 ÷ 3 =**

11) **4,680 ÷ 15 =**

12) **8,936 ÷ 25 =**

Name: Date: Score: Unstoppable Owl

82) Mixed Review

Directions: *Solve each problem and identify any remainders.*

1) **245 × 91 =**

2) **46 × 83 =**

3) **5,736 × 14 =**

4) **7 × 2,498 =**

5) **34,027 × 25 =**

6) **86 × 362 =**

7) **719 ÷ 6 =**

8) **5,782 ÷ 12 =**

9) **946 ÷ 5 =**

10) **7,616 ÷ 8 =**

11) **1,786 ÷ 21 =**

12) **64,718 ÷ 50 =**

Name: Date: Score: Unstoppable Owl

83) Mixed Review

Directions: *Solve each problem and identify any remainders.*

1) **9 × 3,476 =**

2) **48 × 65 =**

3) **875 × 6 =**

4) **2,000 × 73 =**

5) **26 × 289 =**

6) **5 × 17,848 =**

7) **6,485 ÷ 30 =**

8) **2,792 ÷ 4 =**

9) **834 ÷ 15 =**

10) **5,178 ÷ 7 =**

11) **1,265 ÷ 11 =**

12) **7,658 ÷ 25 =**

Name: Date: Score: Unstoppable Owl

84) Mixed Review

Directions: *Solve each problem and identify any remainders.*

1) **17 × 18 =**

2) **54 × 99 =**

3) **6,352 × 4 =**

4) **51 × 38 =**

5) **746 × 35 =**

6) **3 × 5,628 =**

7) **700 ÷ 12 =**

8) **3,725 ÷ 9 =**

9) **14,758 ÷ 30 =**

10) **1,637 ÷ 5 =**

11) **2,334 ÷ 250 =**

12) **371 ÷ 15 =**

Name: Date: Score: Unstoppable Owl

85) Mixed Review

Directions: *Solve each problem and identify any remainders.*

1) **2,861 × 5 =**

2) **486 × 37 =**

3) **94 × 126 =**

4) **7 × 460 =**

5) **2,197 × 14 =**

6) **53 × 28 =**

7) **5,723 ÷ 11 =**

8) **715 ÷ 3 =**

9) **2,683 ÷ 6 =**

10) **9,517 ÷ 15 =**

11) **764 ÷ 20 =**

12) **1,762 ÷ 40 =**

Name: Date: Score:

Unstoppable Owl

86) Mixed Review

Directions: *Solve each problem and identify any remainders.*

1) **45 × 75 =**

2) **6 × 539 =**

3) **4 × 6,052 =**

4) **29 × 37 =**

5) **120 × 243 =**

6) **27 × 428 =**

7) **23,706 ÷ 50 =**

8) **315 ÷ 4 =**

9) **250 ÷ 9 =**

10) **638 ÷ 12 =**

11) **7,000 ÷ 150 =**

12) **437 ÷ 5 =**

Name: Date: Score: Unstoppable Owl

87) Mixed Review

Directions: *Solve each problem and identify any remainders.*

1) **46 × 32 =**

2) **279 × 8 =**

3) **17 × 83 =**

4) **450 × 52 =**

5) **2 × 65,392 =**

6) **7 × 259 =**

7) **653 ÷ 8 =**

8) **1,794 ÷ 15 =**

9) **278 ÷ 11 =**

10) **4,375 ÷ 25 =**

11) **5,637 ÷ 200 =**

12) **4,832 ÷ 7 =**

Name: Date: Score: Unstoppable Owl

88) Mixed Review

Directions: *Solve each problem and identify any remainders.*

1) $4 \times 13{,}520 =$

2) $72 \times 35 =$

3) $41 \times 286 =$

4) $5 \times 4{,}863 =$

5) $30 \times 652 =$

6) $7{,}129 \times 6 =$

7) $15{,}846 \div 2 =$

8) $319 \div 5 =$

9) $445 \div 12 =$

10) $3{,}264 \div 25 =$

11) $3{,}728 \div 4 =$

12) $7{,}842 \div 30 =$

Name: Date: Score: *Unstoppable Owl*

89) Mixed Review

Directions: *Solve each problem and identify any remainders.*

1) **2,651 × 9 =**

2) **64 × 28 =**

3) **438 × 36 =**

4) **2 × 14,793 =**

5) **56 × 227 =**

6) **6,592 × 3 =**

7) **6,675 ÷ 11 =**

8) **11,556 ÷ 25 =**

9) **581 ÷ 7 =**

10) **2,168 ÷ 50 =**

11) **624 ÷ 12 =**

12) **781 ÷ 5 =**

Name: Date: Score:

Unstoppable Owl

90) Mixed Review

Directions: *Solve each problem and identify any remainders.*

1) $46 \times 37 =$

2) $23 \times 519 =$

3) $3{,}725 \times 6 =$

4) $5 \times 8{,}184 =$

5) $427 \times 32 =$

6) $1{,}279 \times 4 =$

7) $5{,}232 \div 6 =$

8) $3{,}472 \div 20 =$

9) $1{,}366 \div 15 =$

10) $380 \div 9 =$

11) $752 \div 12 =$

12) $2{,}783 \div 250 =$

Name: Date: Score: Unstoppable Owl

91) Word Problems

Knowing ***how*** to multiply and divide are important skills, but even more important is knowing ***when*** to multiply or divide.

The word problems in this section are designed to portray real-world situations. Pay close attention to the wording of these problems. They often contain **keywords** that may help to indicate whether multiplication or division is needed to solve the problem.

Common Multiplication Keywords:

- Total
- Times
- Altogether
- In all
- Double
- Triple
- By
- Every

Common Division Keywords:

- Per
- Equal
- Divided
- Each
- Separated
- Parts
- Evenly
- Shared

1) **Alex owns a pet store. The store has 5 fish tanks and 475 goldfish. If Alex puts an equal number of goldfish in each tank, how many goldfish are in each tank?**

2) **The pet store uses 28 scoops of fish food each day to feed the goldfish. At this rate, how many scoops of fish food are used each week?**

3) **13 boxes were delivered to the store. Each box contained 12 bags of dog food. How many bags of dog food is that altogether?**

4) **Alex has 84 dog toys to place on 3 shelves. If he wants to put an even number of dog toys on each shelf, how many should he put on each shelf?**

5) **Alex has 200 crickets to feed to the lizards. If he gives an equal number of crickets to each of the 8 lizards, how many crickets will each lizard get?**

6) **Alex is cutting grapes into small pieces to feed to the parrots. If the pet store has 36 parrots, and Alex wants to give each parrot 25 pieces, how many pieces will Alex need to cut?**

Name: Date: Score: *Unstoppable Owl*

92) Word Problems

1) **Tina owns an ice-cream shop. Today, she sold 286 scoops of strawberry ice-cream and 4 times as many scoops of vanilla ice-cream. How many scoops of vanilla ice-cream were sold?**

2) **The ice-cream cones come in stacks of 24 cones. Tina currently has 65 of these stacks. How many ice-cream cones does Tina have?**

3) **Tina has 8 full buckets of chocolate ice-cream. Each of these buckets serves 157 scoops. How many scoops is that altogether?**

4) **The ice-cream shop has 27 tables and 162 chairs. If Tina puts an equal number of chairs around each table, how many chairs will each table have?**

5) **Tina has 9,000 napkins that she is storing in 12 boxes. If there is an equal number of napkins in each box, how many napkins does each box contain?**

6) **Tina wants to place 25 napkins on each of her 27 tables for customers to use. How many napkins will that require?**

7) **A bag of chocolate chips contains 600 chips. If Tina uses 8 chips to make each chocolate chip sundae, how many chocolate chip sundaes can be made with one bag of chocolate chips?**

8) **If each bag of chocolate chips contains 600 chips, and Tina has 39 bags of chocolate chips, how many chocolate chips does she have altogether?**

Name: Date: Score: Unstoppable Owl

93) Word Problems

1) Jake owns an auto shop. His shop sold 387 sets of tires in July. If each set of tires contains 4 tires, how many tires were sold in July?

2) Jake spent a total of 468 minutes placing a total of 52 tires on vehicles today. On average, how many minutes did it take him to place each tire?

3) It took Jake 468 minutes to place the tires. How long is that when the time is stated in hours and minutes? (There are 60 minutes in an hour.)

4) 5 lug nuts are needed to attach each tire. Knowing this, how many lug nuts did Jake use to attach the 52 tires today?

5) A box contains 1,280 lug nuts. If it takes 5 lug nuts to attach each tire, how many tires can Jake attach using the lug nuts in this box?

6) Jake has 6 boxes of lug nuts. If each box has 1,280 lug nuts, how many lug nuts does Jake have in these 6 boxes combined?

7) It takes Jake 15 minutes to change a vehicle's oil, but it takes him 360 minutes to change a vehicle's transmission. How many times longer does it take to change the transmission than the oil?

8) It takes Jake 45 minutes to change a vehicle's brake pads. Knowing this, how many minutes should it take him to change the brake pads for 7 vehicles?

Name: Date: Score: Unstoppable Owl

94) Word Problems

1) **Ben is a school principal. There are 568 students in the school, and he would like the school to give each student 3 pencils. How many pencils is that altogether?**

2) **Ben orders 1,600 more pencils for the school. They arrive in 25 boxes. If each box contains the same number of pencils, how many pencils are in each box?**

3) **Ben's school has 27 classrooms. He plans to evenly divide the school's 3,100 markers amongst the classrooms. Any leftovers he will save for next year. How many markers should each classroom get?**

4) **Ben also wants each of the 27 classrooms to receive 3,400 sheets paper. How many sheets paper is that altogether?**

5) **Ben realizes he will need to place an order for more paper. He wants to order 130,000 sheets of paper. If the paper is sold in packages of 250 sheets. How many packages should he order?**

6) **Buses are needed for a school field trip. If each bus can seat 72 students, how many buses will be needed to transport the 568 students?**

7) **Claire is a teacher. She has 23 students in her class. If each student reads 48 books throughout the school year, how many books will the class have read altogether?**

8) **So far, Claire's 23 students have read 19,504 pages this year. How many pages is that per student? (*Round your answer to the nearest whole number.*)**

Name: Date: Score: Unstoppable Owl

95) Word Problems

1) **Mary is the city gardener. She ordered 2,800 tulip bulbs to plant. They arrived in packages that had 16 bulbs in each package. How many packages were there?**

2) **Of the 2,800 tulip bulbs Mary planted, a fifth were red tulip bulbs. How many red tulips bulbs did she plant?**

3) **There are 165 bushes for Mary to trim. If it takes her 6 minutes to trim each bush, how many minutes will it take to trim all 165 bushes?**

4) **Mary estimates that there are 25 roses on each of the 189 rose bushes. How many roses is that altogether?**

5) **The city estimates that 48,600 people visited the gardens last month. If there were 30 days last month, on average, how many people visited the gardens each day.**

6) **Mary planted 24 rows of marigolds. If there were 64 marigolds in each row, how many marigolds did Mary plant?**

7) **Mary planted 627 lilies and three times as many daffodils. How many daffodils did she plant?**

8) **Mary planted sunflowers in groups of 8 sunflowers. If she planted a total of 1,160 sunflowers, how many groups of sunflowers were there?**

Name: Date: Score: Unstoppable Owl

96) Word Problems

1) Kyle owns a bakery. The oven can bake a batch of 250 muffins in 23 minutes. If Kyle wants to bake 3,500 muffins, how long will it take to bake them all? (State the answer in hours and minutes.)

2) Of the 3,500 muffins baked, a 7th of them are banana nut muffins and half of the banana nut muffins have chocolate chunks in them. How many of the muffins are banana nut with chocolate chunks?

3) 8 containers of blueberries were used to make 1,250 blueberry muffins. If each container of blueberries holds 2,500 blueberries, on average, how many blueberries are in each of the blueberry muffins?

4) Kyle baked 7 batches of cupcakes, then placed these cupcakes in even amounts on 4 shelves. If there were 300 cupcakes in each batch, how many cupcakes did Kyle put on each shelf?

5) A batch of cranberry cupcakes uses 1,800 cranberries, but Kyle only wants to make a half batch. If there are 60 cranberries in each container, how many containers of cranberries will Kyle need to make half a batch of cranberry muffins?

6) Kyle received an order for 54 vanilla cupcakes and three times as many chocolate cupcakes, but only half of the chocolate cupcakes should be frosted. How many chocolate cupcakes should Kyle frost?

7) Kyle made 2,520 brownie bars. A fourth of them were topped with caramel, and a third of those caramel topped brownie bars had walnut pieces added to them. How many of the bars are topped with caramel and have walnut pieces?

8) Kyle sold 51 pies and a third of them were French silk pies. If each French silk pie contained 6 slices, how many slices of French silk pie were sold altogether?

Name: Date: Score: Unstoppable Owl

97) Word Problems

1) Amy has a farm. The 2,485 chickens each lay about 6 eggs a week. On average, how many eggs does her flock of chickens lay each day?

2) After collecting 10,200 eggs, Amy placed them into egg cartons that each hold 12 eggs. Then she loaded an equal amount of egg cartons onto 5 shipping crates. How many egg cartons were on each shipping crate?

3) Amy had a herd of 278 cows, but the herd has since grown to triple that number. Now, Amy keeps half the herd in one pasture and half the herd in a second pasture. How many cows does she keep in each pasture?

4) Amy's trailer can haul 165 hay bales in one full load. If she transported 15 loads of hay bales, then stacked the bales into 9 even piles, how many hay bales are in each pile.

5) Each day, Amy feeds one hay bale for every 8 sheep in the herd. If she has 184 sheep in her herd, how many hay bales will she feed them in a week?

6) It takes Amy 5 minutes to sheer each of the 184 sheep. How many hours and minutes will it take her to sheer the whole herd?

7) Each of Amy's 184 sheep produce about 4 kg of fleece a year. Amy sells the fleece in 16 kg bundles. How many 16 kg bundles of fleece does her sheep herd produce?

8) A customer bought 112 kg of fleece and told Amy that it takes 2 kg of fleece to make 3 sweaters. How many sweaters can be made with the fleece this customer bought?

Name: Date: Score: Unstoppable Owl

98) Word Problems

1) Pete builds houses. Each shingle he installs requires 3 nails. If he installed 57 bundles of shingles and each bundle contains 48 shingles, how many nails did Pete use?

2) Pete had 27 boxes of screws. Each box contained 1,250 screws. Pete then divided these screws evenly between the 5 workers helping to build the house. How many screws did each worker get?

3) Pete had 627 boards, then each board was cut into 5 pieces. A third of the pieces are long pieces. How many long pieces are there?

4) It takes 85 minutes to install a window. If there are 15 windows to install, how much time, stated in hours and minutes, will it take to install all the windows?

5) Pete took a fifth of the 2,090 short board pieces and drilled 8 holes into each of them. How many holes did he drill?

6) Pete bought 6,480 floor tiles. They came in boxes that had 30 tiles in each box. A fourth of the boxes had brown floor tiles. How many boxes had brown floor tiles?

7) The house Pete built last month needed 13,200 bricks. The current house being built only needs half that amount. If Pete can haul 300 bricks in one full load, how many full loads will it take to haul enough bricks for the current house?

8) Pete's mixer can mix 2 bags of concrete every 5 minutes. At this rate, how long will it take him to mix 180 bags of concrete? (State your answer in hours, using a decimal if needed.)

Name: Date: Score: Unstoppable Owl

99) Word Problems

1) Kelly is a ranger at a nature preserve. She says a squirrel hides about 25 nuts each hour. If a squirrel hides nuts 9 hours a day for a week, how many nuts will it hide.

2) Kelly says a squirrel may hide 9,000 nuts a year, but a fourth of them are stolen. She estimates that a fifth of the stolen nuts are eaten by bears. If true, how many nuts out of 9,000 will bears eat?

3) Kelly says there are twice as many swan nests this year than there were 10 years ago, and each nest will have an average of 6 eggs. If there were 382 swan nests 10 years ago, how many swan eggs are expected this year?

4) Kelly said that 7,644 bats live in the cave, and that half of these bats are brown bats. Kelly says each brown bat is believed to eat 2,000 bugs each night. If true, how many bugs did the brown bats from the cave eat last night?

5) On average 46 salmon have been seen swimming upriver every 10 minutes. At this rate, how many salmon would be seen swimming upriver during an 8-hour period?

6) Kelly says there is an estimated 194,000 salmon in the river. Half of them are female, and each female will lay about 1,200 eggs. How many salmon eggs are expected to be laid?

7) For every 100,000,000 salmon eggs laid, only a fifth of them are expected to hatch into fry. A tenth of the fry are expected to reach the ocean. How many fry are expected to reach the ocean?

8) 28,000 people visited the preserve last month, but only half of them went hiking, and just an eighth of the hikers made it to the hidden waterfall. How many people made it to the hidden waterfall?

Name: Date: Score: Unstoppable Owl

100) Word Problems

1) John owns a restaurant. He served 548 French toast meals and 3 times as many pancake platters. If each pancake platter comes with 4 pancakes, how many pancakes did he serve?

2) John made 1,096 pieces of French toast today. If 3 eggs are needed to make 4 pieces of French toast, how many eggs were needed to make the French toast today?

3) John ordered 4,050 eggs. They arrived in cartons that each hold 30 eggs. An equal number of these cartons were placed onto 5 shelves. How many cartons are on each shelf?

4) John made 2,148 pancakes on Saturday. A sixth of these were chocolate chip pancakes. Each chocolate chip pancake is made with 15 chocolate chips. How many chocolate chips were used to make the chocolate chip pancakes?

5) A fourth of the 2,148 pancakes were served with a fruit topping and a third of those pancakes that were served with a fruit topping had strawberries as the fruit. How many pancakes were served with a strawberry fruit topping?

6) A third of John's customers drink coffee with their meal. On average, each of these customers drink 2 cups of coffee. If 4,689 customers came to the restaurant on Saturday, how many cups of coffee were served that day?

7) John can make 20 pancakes every 5 minutes. He then places the pancakes in stacks of 4. At this rate, how long will it take him to make 85 stacks of pancakes

8) The dish washer can wash 48 plates at a time. If a load of plates takes 4 minutes to wash, how many plates can be washed In an hour?

101) Answer Key

Page 1:
1) 150 2) 240 3) 1,600 4) 2,700
5) 12,000 6) 28,000 7) 300 8) 800
9) 2,000 10) 6,000 11) 10,000 12) 420

Page 2:
1) 4,800 2) 210 3) 18,000 4) 1,600
5) 120,000 6) 600 7) 4,000 8) 36,000
9) 7,200 10) 140 11) 100,000 12) 6,300
13) 400 14) 8,000 15) 4,900 16) 1,600
17) 90,000 18) 1,800 19) 81,000 20) 320,000

Page 3:
1) 4,200 2) 6,000 3) 320,000 4) 140,000
5) 300,000 6) 200,000 7) 10,000 8) 1,200,000
9) 16,000 10) 900 11) 210,000 12) 60,000

Page 4:
1) 150,000 2) 28,000 3) 1,000,000 4) 800
5) 540,000 6) 400,000 7) 36,000 8) 4,000,000
9) 5,600 10) 30,000 11) 1,200,000 12) 250,000
13) 10,000 14) 4,000 15) 240,000 16) 600
17) 48,000 18) 21,000 19) 3,000 20) 36,000,000

Page 5:
1) 360 2) 273 3) 1,084 4) 3,160
5) 747 6) 1,362 7) 7,336 8) 18,506

Page 6:
1) 118 2) 288 3) 1,626 4) 3,808
5) 425 6) 5,436 7) 7,760 8) 22,572

Page 7:
1) 72 2) 2,608 3) 36,980 4) 4,067
5) 178 6) 39,732 7) 3,663 8) 90
9) 595 10) 3,168 11) 1,689 12) 168
13) 30,200 14) 128 15) 590 16) 342
17) 45 18) 38,465 19) 405 20) 2,019

Page 8:
1) 351 2) 465 3) 29,232 4) 5,472
5) 3,003 6) 4,088 7) 1,478 8) 1,518
9) 7,512 10) 1,743 11) 5,056 12) 35,780
13) 360 14) 684 15) 188 16) 256
17) 86,292 18) 138,096 19) 188,450 20) 37,650

Page 9:
1) 2,952 2) 1,050 3) 59,604 4) 40,696
5) 172 6) 553 7) 387 8) 183
9) 1,032 10) 19,500 11) 7,776 12) 23,541
13) 623 14) 1,256 15) 52,308 16) 380
17) 473,538 18) 31,716 19) 1,083,775 20) 65,592

Page 10:
1) 4,105 2) 468 3) 51,513 4) 12,816
5) 324 6) 891 7) 304 8) 464
9) 2,594 10) 5,157 11) 37,448 12) 4,361
13) 3,790 14) 195 15) 114 16) 26,757
17) 3,611,032 18) 589,561 19) 1,082,004 20) 46,800

Page 11:
1) 32,795 2) 2,635 3) 376 4) 7,284
5) 1,470 6) 11,794 7) 18,486 8) 1,158
9) 2,912 10) 588 11) 2,145 12) 74,712
13) 222 14) 252 15) 222 16) 330
17) 335,255 18) 6,023,472 19) 50,276 20) 70,488

Page 12:
1) 360 2) 273 3) 1,084 4) 3,160
5) 747 6) 1,362 7) 7,336 8) 18,506

Page 13:
1) 3,682 2) 1,408 3) 632 4) 1,917
5) 31,375 6) 4,140 7) 4,902 8) 16,318
9) 371 10) 1,863 11) 760 12) 292
13) 29,056 14) 28,310 15) 161,588 16) 4,248

Page 14:
1) 23,810 2) 5,104 3) 642 4) 11,952
5) 476 6) 4,239 7) 1,970 8) 2,220
9) 14,832 10) 2,382 11) 688 12) 1,745
13) 1,052,084 14) 167,276 15) 90,902 16) 34,416

Page 15:
1) 2,700 2) 14,586 3) 7,336 4) 13,796
5) 315 6) 365 7) 2,912 8) 738
9) 1,518 10) 53,800 11) 2,860 12) 5,838
13) 35,586 14) 112,428 15) 850,698 16) 21,455

Page 16:
1) 268 2) 406 3) 657 4) 420
5) 17,463 6) 3,776 7) 1,788 8) 55,518
9) 1,854 10) 3,645 11) 15,564 12) 456
13) 234,135 14) 25,928 15) 52,112 16) 15,984

Page 17:
1) 736 2) 315 3) 595 4) 690
5) 5,220 6) 801 7) 9,954 8) 11,148
9) 185 10) 3,896 11) 13,706 12) 52,770
13) 703,934 14) 172,167 15) 374,067 16) 38,088

Page 18:
1) 11,034 2) 2,664 3) 63,315 4) 2,244
5) 240 6) 602 7) 368 8) 624
9) 24,348 10) 3,944 11) 1,508 12) 48,834
13) 210,455 14) 2,245,708 15) 306,000 16) 28,416

Page 19:
1) 891 2) 140 3) 133 4) 138
5) 1,544 6)2,940 7) 17,230 8) 35,304
9) 66,500 10) 1,872 11) 35,613 12) 680
13) 73,872 14) 130,744 15) 302,868 16) 54,320

Page 20:
1) 498 2) 154 3) 288 4) 135
5) 3,395 6) 19,744 7) 2,056 8) 3,405
9) 1,368 10) 33,968 11) 833 12) 56,835
13) 421,984 14) 57,695 15) 655,902 16) 34,284

Page 21:
1) 1,824 2) 2,660 3) 1,406 4) 2,046
5) 2,378 6) 2,380 7) 378 8) 5,525

Page 22:
1) 4,836 2) 2,183 3) 4,455 4) 560
5) 4,278 6) 950 7) 5,494 8) 3,977
9) 3,078 10) 903 11) 195 12) 1,885
13) 3,200 14) 312 15) 1,292 16) 8,316

Page 23:
1) 2,257 2) 1,260 3) 1,344 4) 3,872
5) 945 6) 456 7) 2,544 8) 6,750
9) 2,047 10) 1,386 11) 3,276 12) 442
13) 4,370 14) 2,613 15) 4,455 16) 5,520

Page 24:
1) 1,161 2) 7,826 3) 221 4) 1,188
5) 3,572 6) 5,070 7) 832 8) 3,150
9) 2,352 10) 2,886 11) 600 12) 588
13) 8,633 14) 1,323 15) 828 16) 1,872

Page 25:
1) 2,679 2) 308 3) 7,068 4) 646
5) 1,785 6) 1,666 7) 4,672 8) 4,263
9) 1,512 10) 256 11) 1,702 12) 3,825
13) 378 14) 7,050 15) 4,928 16) 888

Page 26:
1) 2,835 2) 2,646 3) 868 4) 864
5) 5,795 6) 2,640 7) 3,024 8) 323
9) 2,225 10) 1,554 11) 1,210 12) 7,252
13) 1,264 14) 4,644 15) 1,377 16) 2,832

Page 27:
1) 3,680 2) 1,736 3) 6,952 4) 1,679
5) 3,654 6) 1,110 7) 1,032 8) 1,530
9) 7,144 10) 1,062 11) 5,135 12) 7,470
13) 910 14) 1,428 15) 5,952 16) 3,240

Page 28:
1) 52,041 2) 10,619 3) 34,960 4) 4,176
5) 15,012 6) 34,447 7) 6,912 8) 82,128

Page 29:
1) 90,525 2) 397,020 3) 97,722
4) 107,374

Page 30:
1) 167,079 2) 135,150 3) 40,194
4) 32,200 5) 414,200 6) 394,776
7) 164,500 8) 8,245 9) 18,738
10) 324,285 11) 336,766 12) 215,305

Page 31:
1) 59,006 2) 309,568 3) 20,102
4) 322,560 5) 586,365 6) 12,792
7) 49,284 8) 255,375 9) 113,770
10) 1,393,446 11) 154,380 12) 34,560

Page 32:
1) 114,540 2) 19,807 3) 31,382
4) 305,578 5) 76,704 6) 198,975
7) 24,853 8) 12,393 9) 350,658
10) 72,998 11) 613,955 12) 138,945

Page 33:
1) 2,505,768 2) 42,194 3) 24,700
4) 58,904 5) 133,416 6) 270,725
7) 8,502 8) 252,200 9) 260,268
10) 206,062 11) 323,610 12) 8,256

Page 34:
1) 179,587 2) 63,504 3) 253,154
4) 11,526 5) 198,575 6) 431,970
7) 204,570 8) 40,110 9) 375,964
10) 365,415 11) 767,576 12) 29,750

Page 35:
1) 242,060 2) 106,266 3) 339,740
4) 19,604 5) 275,424 6) 151,872
7) 275,328 8) 326,678 9) 975,186
10) 497,472 11) 676,396 12) 9,184

Page 36:
1) 15,124 2) 3,240 3) 151,188
4) 265,280 5) 37,221 6) 58,875
7) 129,600 8) 18,702 9) 94,080
10) 229,600 11) 923,186 12) 6,080

Page 37:
1) 76,464 2) 36,577 3) 112,578
4) 57,712 5) 201,892 6) 3,312
7) 13,005 8) 74,140 9) 448,335
10) 275,142 11) 197,502 12) 13,412

102) Answer Key

Page 38:
1) 290,400 2) 21,560 3) 167,382 4) 496,984
5) 447,105 6) 2,574 7) 183,816 8) 164,582
9) 431,568 10) 802,240 11) 345,655 12) 10,752

Page 39:
1) 125,936 2) 128,700 3) 143,451 4) 6,688
5) 137,242 6) 956,388 7) 20,412 8) 288
9) 199,758 10) 127,368 11) 938,862 12) 6,237

Page 40:
1) 2,242 2) 101,880 3) 175,476 4) 43,078
5) 197,372 6) 8,703 7) 34,540 8) 11,358
9) 568,100 10) 329,594 11) 174,175 12) 36,668

Page 41:
1) 29 2) 42 3) 17 4) 18

Page 42:
1) 43 2) 125 3) 28 4) 56

Page 43:
1) 34 2) 257 3) 81 4) 183
5) 23 6) 143 7) 59 8) 191
9) 1,758 10) 1,072 11) 2,282 12) 215

Page 44:
1) 47 2) 637 3) 214 4) 291
5) 36 6) 461 7) 93 8) 117
9) 22,937 10) 3,508 11) 1,587 12) 53

Page 45:
1) 409 2) 52 3) 735 4) 89
5) 29 6) 670 7) 362 8) 243
9) 398 10) 12,602 11) 3,972 12) 750

Page 46:
1) 573 2) 99 3) 312 4) 3,048
5) 293 6) 719 7) 3,754 8) 76
9) 5,168 10) 2,370 11) 33,749 12) 83

Page 47:
1) 340 2) 791 3) 542 4) 956
5) 57 6) 716 7) 173 8) 444
9) 3,951 10) 8,530 11) 2,749 12) 178

Page 48:
1) 468 2) 92 3) 18 4) 2,604
5) 226 6) 968 7) 635 8) 2,842
9) 4,500 10) 1,250 11) 25,105 12) 644

Page 49:
1) 293 2) 332 3) 182 4) 2,381
5) 141 6) 2,175 7) 3,085 8) 722
9) 3,217 10) 6,082 11) 3,726 12) 663

Page 50:
1) 252 2) 37 3) 1,753 4) 940
5) 143 6) 298 7) 1,671 8) 625
9) 7,169 10) 1,750 11) 8,395 12) 764

Page 51:
1) 7 apples, 1 remainder
2) 4 coins, 3 remainders
3) 3 cookies, no remainders
4) 6 carrots, 2 remainders

Page 52:
1) 37 R1 2) 58 R2 3) 136 R4 4) 220R5

Page 53:
1) 16 R4 2) 58 R2 3) 1,256 R2 4) 15 R8
5) 131 R5 6) 1,609 R1 7) 1,385 R2 8) 461 R1
9) 3,571 R3 10) 1,568 R4 11) 782 R8 12) 76 R4

Page 54:
1) 124 R3 2) 346 R2 3) 45 R1 4) 322 R5
5) 189 6) 635 R4 7) 973 R4 8) 222 R2
9) 4,942 R1 10) 1,909 R3 11) 1,958 12) 633 R2

Page 55:
1) 944 R6 2) 91 R5 3) 1,333 R2 4) 125 R4
5) 1,355 R1 6) 472 R2 7) 124 R 7 8) 34 R 1
9) 3,322 R1 10) 8,472 R5 11) 1,860 12) 944 R2

Page 56:
1) 1,432 R2 2) 338 R1 3) 428 R4 4) 214 R1
5) 252 6) 351 R7 7) 67 R3 8) 818 R3
9) 2,715 R4 10) 2,983 R2 11) 2,240 R2 12) 2,972 R2

Page 57:
1) 581 R2 2) 44 R4 3) 439 R2 4) 94 R7
5) 1,264 R6 6) 156 R1 7) 250 8) 975 R4
9) 12,368 R2 10) 1,956 R1 11) 766 R2 12) 678 R5

Page 58:
1) 808 R3 2) 935 R1 3) 126 4) 336 R2
5) 780 R4 6) 383 R3 7) 158 R4 8) 662 R3
9) 15,850 R2 10) 8,354 R2 11) 88 R2 12) 770 R1

Page 59:
1) 556 R4 2) 182 R2 3) 140 R2 4) 2,327 R2
5) 917 R3 6) 428 R2 7) 319 R1 8) 650
9) 3,211 R3 10) 946 R7 11) 9,128 R1 12) 1,518 R2

Page 60:
1) 159 R2 2) 446 R4 3) 170 R5 4) 845 R2
5) 213 R1 6) 2,520 R1 7) 374 R2 8) 75 R5
9) 894 R3 10) 7,263 R2 11) 2,263 R1 12) 987 R2

Page 61:
1) 14.6 2) 117.5 3) 159.5 4) 74.8

Page 62:
1) 28.5 2) 22.5 3) 639.2 4) 1,982.5
5) 167.8 6) 13.5 7) 862.5 8) 689.4
9) 2,011.6 10) 1,799.5 11) 15,254.2 12) 3,419.5

Page 63:
1) 20.25 2) 2.75 3) 39.625 4) 31.75

Page 64:
1) 20.625 2) 72.75 3) 57.5 4) 168.0
5) 34.25 6) 32.875 7) 87.5 8) 68.75
9) 12.5 10) 0.75 11) 78.125 12) 27.75

Page 65:
1) 7.38 2) 25.07 3) 1.95 4) 214.33
5) 5.67 6) 8.72 7) 40.78 8) 20
9) 143.28 10) 10.25 11) 0.4 12) 637.81
13) 8.22 14) 24.17 15) 29.33 16) 2.13

Page 66:
1) 44.38 2) 64.89 3) 70.29 4) 92.67
5) 5.78 6) 14.33 7) 44.17 8) 162.56
9) 15.13 10) 78.33 11) 27.78 12) 49.83

Page 67:
1) 81.78 2) 78.43 3) 15.83 4) 54.67
5) 35.29 6) 85.17 7) 1.63 8) 3.86
9) 12.5 10) 252.67 11) 40.88 12) 0.56

Page 68:
1) 94.67 2) 71.38 3) 4.11 4) 98.29
5) 9.83 6) 21.86 7) 42.89 8) 11.75
9) 286.33 10) 23.78 11) 73.63 12) 23.67

Page 69:
1) 63.57 2) 8.14 3) 2.89 4) 32.56
5) 13.67 6) 158.67 7) 9.63 8) 3.57
9) 49.88 10) 29.17 11) 64.75 12) 68.83

Page 70:
1) 54.83 2) 0.38 3) 19.67 4) 7.88
5) 97.14 6) 83.67 7) 34.75 8) 5.38
9) 497.43 10) 94.6 11) 54.22 12) 20.63

Page 71:
1) 36 2) 148 3) 35 4) 86
5) 65 6) 29 7) 157 8) 231

Page 72:
1) 462 2) 243 3) 411 4) 33
5) 526 6) 520 7) 24 8) 38
9) 143 10) 382 11) 821 12) 423

Page 73:
1) 304 2) 21 3) 215 4) 342
5) 643 6) 203 7) 412 8) 652
9) 729 10) 1,431 11) 2,434 12) 2,754

Page 74:
1) 42 2) 679 3) 524 4) 748
5) 34 6) 240 7) 4,135 8) 1,275
9) 867 10) 2,403 11) 431 12) 534

Page 75:
1) 542 2) 648 3) 3,412 4) 257
5) 324 6) 938 7) 753 8) 964
9) 729 10) 4,113 11) 45 12) 424

Page 76:
1) 41 R12 2) 208 R5 3) 81 R3 4) 192 R10
5) 27 R16 6) 57 R3 7) 32 R9 8) 151 R20

Page 77:
1) 428 R3 2) 554 R22 3) 2,368 R12 4) 39 R8
5) 309 R5 6) 359 R8 7) 4,075 R3 8) 856 R39
9) 862 R5 10) 532 R6 11) 941 R9 12) 72 R12

Page 78:
1) 702 R16 2) 338 R2 3) 214 R2 4) 544 R18
5) 409 R7 6) 352 R13 7) 541 R4 8) 28 R12
9) 71 R9 10) 138 R6 11) 723 R4 12) 343 R11

Page 79:
1) 316 R7 2) 31 R59 3) 416 R8 4) 327 R20
5) 62 R2 6) 241 R155 7) 131 R7 8) 259 R34
9) 56 R37 10) 127 R30 11) 91 R16 12) 32 R4

Page 80:
1) 142 R18 2) 31 R97 3) 148 R2 4) 642 R24
5) 465 R23 6) 41 R8 7) 287 R18 8) 28 R183
9) 1,782 R6 10) 9 R56 11) 73 12) 189 R18

Page 81:
1) 4,216 2) 2,790 3) 33,320 4) 52,392
5) 4,495 6) 2,680 7) 188 R2 8) 535
9) 179 R2 10) 324 R1 11) 312 12) 357 R11

Page 82:
1) 22,295 2) 3,818 3) 80,304 4) 17,486
5) 850,675 6) 31,132 7) 119 R5 8) 481 R10
9) 189 R1 10) 952 11) 85 R1 12) 1,294R18

Page 83:
1) 31,284 2) 3,120 3) 5,250 4) 146,000
5) 7,514 6) 89,240 7) 216 R5 8) 698
9) 55 R9 10) 739 R5 11) 115 12) 306 R8

103) Answer Key

Page 84:
1) 306 2) 5,346 3) 25,408 4) 1,938
5) 26,110 6) 16,884 7) 58 R4 8) 413 R8
9) 491 R28 10) 327 R2 11) 9 R84 12) 24 R11

Page 85:
1) 14,305 2) 17,982 3) 11,844 4) 3,220
5) 30,758 6) 1,484 7) 520 R3 8) 238 R1
9) 447 R1 10) 634 R7 11) 38 R4 12) 44 R2

Page 86:
1) 3,375 2) 3,234 3) 24,208 4) 1,073
5) 29,160 6) 11,556 7) 474 R6 8) 78 R3
9) 27 R7 10) 53 R2 11) 46 R100 12) 87 R2

Page 87:
1) 1,472 2) 2,232 3) 1,411 4) 23,400
5) 130,784 6) 1,813 7) 81 R5 8) 119 R9
9) 25 R3 10) 175 11) 28 R37 12) 690 R2

Page 88:
1) 54,080 2) 2,520 3) 11,726 4) 24,315
5) 19,560 6) 42,774 7) 7,923 8) 63 R4
9) 37 R1 10) 130 R14 11) 932 12) 261 R12

Page 89:
1) 23,859 2) 1,792 3) 15,768 4) 29,586
5) 12,712 6) 19,776 7) 606 R9 8) 462 R6
9) 83 10) 43 R18 11) 52 12) 156 R1

Page 90:
1) 1,702 2) 11,937 3) 22,350 4) 40,920
5) 13,664 6) 5,116 7) 872 8) 173 R12
9) 91 R1 10) 42 R2 11) 62 R8 12) 11 R33

Page 91:
1) 95 goldfish 2) 196 scoops
3) 156 bags 4) 28 dog toys
5) 25 crickets 6) 900 pieces

Page 92:
1) 1,144 vanilla scoops 2) 1,560 cones
3) 1,256 scoops 4) 6 chairs per table
5) 750 napkins per box 6) 675 napkins
7) 75 sundaes 8) 23,400 chocolate chips

Page 93:
1) 1,548 tires 2) 9 minutes per tire
3) 7 hours 48 minutes 4) 260 lug nuts
5) 256 tires 6) 7,680 lug nuts
7) 24 times longer 8) 315 minutes

Page 94:
1) 1,704 pencils 2) 64 pencils per box
3) 114 markers 4) 91,800 sheets
5) 520 packages 6) 8 buses (7 is not enough)
7) 1,104 books 8) 848 pages per student

Page 95:
1) 175 packages 2) 560 red tulip bulbs
3) 990 minutes 4) 4,725 roses
5) 1,620 people per day 6) 1,536 marigolds
7) 1,881 daffodils 8) 145 groups

Page 96:
1) 5 hr 22 min (3,500 ÷ 250 × 23 ÷ 60 = 5 R22)
2) 250 muffins (3,500 ÷ 7 ÷ 2 = 250)
3) 16 blueberries (8 × 2,500 ÷ 1,250 = 16)
4) 525 cupcakes per shelf (7 × 300 ÷ 4 = 525)
5) 15 containers (1,800 ÷ 2 ÷ 60 = 15)
6) 81 chocolate cupcakes (54 × 3 ÷ 2 = 81)
7) 210 bars (2,520 ÷ 4 ÷ 3 = 210)
8) 102 slices (51 ÷ 3 × 6 = 102)

Page 97:
1) 2,130 eggs per day (2,485 × 6 ÷ 7 = 2,130)
2) 170 egg cartons (10,200 ÷ 12 ÷ 5 = 170)
3) 417 cows per pasture (278 × 3 ÷ 2 = 417)
4) 275 hay bales (165 × 15 ÷ 9 = 275)
5) 161 bales of hay (184 ÷ 8 × 7 = 161)
6) 15 hours 20 minutes (184 × 5 ÷ 60 = 15 R20)
7) 46 bundles (184 × 4 ÷ 16 = 46)
8) 168 sweaters (112 ÷ 2 × 3 = 168)

Page 98:
1) 8,208 nails (57 × 48 × 3 = 8,208)
2) 6,750 screws (27 × 1,250 ÷ 5 = 6,750)
3) 1,045 long pieces (627 × 5 ÷ 3 = 1,045)
4) 21 hr 15 min (85 × 15 ÷ 60 = 21 R15)
5) 3,344 holes (2,090 ÷ 5 × 8 = 3,344)
6) 54 boxes (6,480 ÷ 30 ÷ 4 = 54)
7) 22 full loads (13,200 ÷ 2 ÷ 300 = 22)
8) 7.5 hours (180 ÷ 2 × 5 ÷ 60 = 7.5)

Page 99:
1) 1,575 nuts (25 × 9 × 7 = 1,575)
2) 450 nuts (9,000 ÷ 4 ÷ 5 = 450)
3) 4,584 swan eggs (382 × 2 × 6 = 4,584)
4) 7,644,000 bugs (7,644 ÷ 2 × 2,000 =7,644,000)
5) 2,208 salmon (60 ÷ 10 × 46 × 8 = 2,208)
or (6 × 46 × 8 = 2,208)
6) 116,400,000 eggs
(194,000 ÷ 2 × 1,200 = 116,400,000)
7) 2,000,000 fry
(100,000,000 ÷ 5 ÷ 10 = 2,000,000)
8) 1,750 people (28,000 ÷ 2 ÷ 8 = 1,750)

Page 100:
1) 6,576 pancakes (548 × 3 × 4 = 6,576)
2) 822 eggs (1,096 ÷ 4 × 3 = 822)
3) 27 cartons per shelf (4,050 ÷ 30 ÷ 5 = 27)
4) 5,370 chocolate chips (2,148 ÷ 6 × 15 = 5,370)
5) 179 pancakes (2,148 ÷ 4 ÷ 3 = 179)
6) 3,126 cups (4,689 ÷ 3 × 2 = 3,126)
7) 1 hr 25 min. (85 × 4 ÷ 20 × 5 ÷ 60 = 1 R25)
8) 720 plates (60 ÷ 4 × 48 = 720)

ISBN: 978-1-63578-552-4
Current contact information can be found at:
www.UnstoppableOwl.com www.LibroStudioLLC.com

Made in United States
North Haven, CT
10 November 2026

22106127R00059